For all the children,

Keep learning and never stop growing.

FOR CAREGIVERS & TEACHERS:

If you'd like to expand your conversation after reading this book, use the Internet or other sources, such as the library, to search the following questions in order to learn interesting facts about the creatures mentioned.

1. What is a seedling plant?
2. How do birds make a nest?
3. How deep under snow can squirrels smell food?
4. Do skunks have unlimited spray?
5. Why do queen bees get marked with a spot of color?
6. What happens inside a chrysalis?
7. Why do turkeys have a wattle?
8. Why do rodents like to chew on acorns?
9. How long is a woodpecker's tongue?
10. Do owls have true eyeballs?
11. What is interesting about a frog's vocal cords?
12. What are moth larvae?

Additional research or project ideas:

How can people help preserve and protect wildlife and forests?
What are some ways to cut down on pollution?
What are some alternative products that we can use in everyday life to save trees?

SO SWEET
A Rustle In
The Leaves
Written & Illustrated
by Julie Rockwell

To request permissions contact the author with subject, "Attention: Permissions" at the following email address:

sosweetbooks@gmail.com
www.sosweetbooks.com

Paperback ISBN: 979-8-9865243-1-3
Hardback ISBN: 979-8-9865243-2-0

March 2023

A seed is planted down in the ground,
and the roots sprout out and all around.

Magic happens as a sapling tree grows,
outward and upward and higher it goes.

Arm-like branches reach out toward the sky,
and the tree grows taller and stands up high.

So many creatures live and grow in the trees;
some hide their nests among the leaves.

Hungry baby birds can be heard chirping loudly
while mom feeds them worms ever so proudly.

Squirrels scurry about, up and down the trunk,

and creeping below is a new mother skunk.

Honey bees fly in and out of the hive

working hard to keep their queen alive.

Caterpillars munch leaves and grow quickly in size

become chrysalises, and then beautiful butterflies.

Turkeys roost up high; so it's been told

and mice will move in when it gets cold.

Woodpeckers knock holes into the bark.

Owls search for prey when it's dark.

Frogs croak happily all through the night.

While most predators sleep, moths will take flight.

How sweet it can be
deep within these trees,

listening to creatures as
they rustle the leaves.

So, when you visit a forest, please be kind,
and be sure to leave only footprints behind.

Happy and healthy, trees in the forests will grow,
and the magic of it all can continue to flow.

THE END

Flip the page

For more fun, go back through the illustrations in this book and see if you can find the following:

a turkey wattle
worms that are about to be eaten
munched leaves with holes in them
a butterfly emerging
a sign with an important message
a baby skunk being carried by mom
creatures that croak and hop
rays of sunlight peeking through trees
someone eating a tiny acorn
5 eggs not yet hatched
footprints in the dirt
a moth larvae nest
a queen bee with a red marking

How many creatures can you spot on the page where the tree is full?
Do you remember what they were doing?
(I'll remind you of one, the worker bees were in the hive protecting their queen.)

www.ingramcontent.com/pod-product-compliance
Lightning Source LLC
LaVergne TN
LVHW071238210926
848486LV00007B/298
* 9 7 9 8 9 8 6 5 2 4 3 2 0 *